I0476394

The Smartphone Buyer's Guide 2015

How to Choose the Right Smartphone for Your Needs

Steve Johnson

Contents

Introduction

Since the beginning of time, the world has slowly developed into a global playground for information and accessibility. There is an infinite number of ways human beings use technology to communicate with one another. Various devices connect us to people all across the globe through wireless technology. The thriving advancements of wireless technology have granted us the ability to remain connected to the world.

Wireless communication is a powerful force in present day. Conversely, the influence of this innovative technology was not always as prevalent. Wireless communication has gradually emancipated the modern-day household from the oppression of phone cords and tangled wires.

Because of wireless technology, the rate at which the human race exchanges information is truly mind-blowing. Wireless technology will continue to have a positive influence on the timeliness of important business decisions, the coverage of major world events, and much more.

Through our many wireless devices and gadgets, we have been able to channel tons of wireless technology in the direction that we choose.

Wireless capabilities will allow the human race to thrive and maintain itself efficiently. The most powerful and popular device available on the market today, is the mobile phone. Quartz reported that out of the 7 billion people on earth, there are as much as 6.8 million mobile phone subscriptions worldwide. That is almost one mobile phone for each human being on earth.

For over 42 years, the mobile phone has been in existence, since its creation it has repetitively improved. The mobile phone went from being an exclusive commodity, to a pop culture icon that everyone loves. The first mobile phone was merely a chunky piece of plastic with a large antenna poking out the top. Its oversized buttons, sword-like antenna, and pale gray shell only presented consumers with one feature: wireless phone calls. The modern day mobile phone presents a sleek and stylish design equipped with superior technology. These high-tech devices have changed the way we do business, communicate, and live.

The mobile phone market is lucrative, and is overflowing with an impressive variety of mobile phones, phone accessories, phone service providers, and phone applications. Nevertheless, a market teeming with a vast variety of a singular product can make phone shopping slightly difficult.

The easiest solution would be to buy the phone that currently holds "the popular vote," or the phone that is trendy and new to the public. However, the excitement and gratification felt by purchasing that new phone will die out, and you will be left with a phone and a bill that you don't need. The inspiration to exhaust the maximum usage of your phone will never show up. Shopping by trends will throw you into a never ending cycle of phone upgrades and wasted money. Don't let this to happen to you.

It would be considered most wise to purchase your mobile phone based on your wants, needs, and budget. Resist the temptation to buy that popular, expensive, phone without carefully considering your options. Why? That chic, expensive, state-of-the-art, phone may not be the phone that is suitable for you.

Sustenance of wealth and individuality is important. Individuals that take themselves and their money seriously will make the decision to choose their phone wisely. They understand that their phone must be compatible with their lifestyle and accommodating to their budget. They know that choosing the right phone will give them the highest level of comfort and functionality. However, in order to pick out the right phone, you need the scoop on the phone industry and the many options that you have.

In this book, The Smartphone Buyer's Guide 2015, you will be given the exclusive on how to effortlessly pick out the phone right for you, you will become familiar with the basic capabilities of the modern-day cell phone. You will also discover some crazy, but intriguing facts about the mobile phone and the development of wireless technology. This eBook will be your access pass into the world of the continuously evolving phenomenon. The knowledge you gain from reading this book will help you get the MOST out of your mobile phone.

From Mobile Phone to Mobile Device to Smartphone: The Evolution of the Cell Phone

Mobile phones have been around for over 40 years. The original mobile phone was an oversized, plastic, shell ornamented by ginormous buttons. It possessed minor capabilities and very limited technology. But back then, mobile phones didn't need to look stylish or cool. The technology housed inside that heavy shell made the phone's appearance bearable.

The original mobile phone came equipped with technology that the world has never seen before. Consumers were presented with options that they never had before. Phone calls could now be placed from any location. No more sending messages by carrier pigeon, no more frequent trips to make a call from your car phone, no more waiting nearly half the day to get home and make important phone calls, and no more loud, jiggling pockets running to the pay phone. This revolutionary and sensational technology outshined the unattractive shell that it was in, because the technology that it housed was pure gold.

In contrast, the current mobile phones are nicely designed. These sleek, slim, and stylish devices contain technology that can outmatch some of the most advanced computers on the market today. The new and improved mobile phone is a portable communication system that can do pretty much whatever we program or "tell" it to do. It is sure to be around and evolve into something better in the years to come.

Smartphones come equipped with carefully developed applications that surpass the functionality of the regular mobile phone. Some of the applications programed into the smartphone can even be found on your regular desktop or laptop computer. Applications and features such as a touchscreen, Bluetooth, Microsoft Office, email access, a QWERTY keyboard, a camera, Internet browsing, a media player, and contact management are some of the MOST BASIC features of the modern-day mobile phone. The smartphone can even be synchronized with your other mobile devices, or even your desktop PC. You can even take a trip back to the past (sort of), and use wireless connectivity to receive phone calls through your car. The only difference is that there will be no wires/cords, better coverage, and fast connectivity. You can even place or answer a phone call without even picking up the phone. The future is looking very bright for the smartphone. Yes indeed.

Developing the Concept

It's hard to imagine what communication was like before the invention of the mobile phone. Sure, talking to a person face-to-face was and still is the most natural and effective way to make an announcement. But what about the lonely grandmother that lives in Timbuktu? I'm sure she is longing to hear the announcement, or to hear from anyone for that matter. Or what about the crazy, daredevil brother that took a year-long escapade to the North Pole? I'm sure he is dying for warm and cozy conversation, or really, anything that will keep him warm. Why should distance cancel them out from partaking in a warming and heartening conversation? Wouldn't it be nice if they could be a part of the moment, witnessing every word without even having a physical presence? With all things being considered, it is safe to say that the earliest composers of wireless technology thought about the same questions and scenarios in a similar of different form.

In 1918, The German Railroad System was the first entity to use the technology of "wireless telephony." They conducted the private testing of the first wireless telephone on military trains traveling between Zossen and Berlin. By 1924, the testing of the telephony equipment went public, and the results turned out to be quite impressive. The success of the testing inspired and founded Zugtelephonie A.G., a company dedicated to supplying telephony equipment entirely for trains. They began the task of installing telephony equipment on all trains traveling between Berlin and Hamburg in 1925. The use of this wireless phone was limited only to the passengers traveling in 1st class.

Decades later, the United States began to develop a system that would allow the wireless telephone to be available for use in the automobile. As a result, this caused the mobile phone service to be inaugurated on June 17, 1946. This encouraged AT&T Mobile Telephone Service, to become one of the first mobile phone carriers in the United States. Shortly after, a slew of companies arose from the shadows to offer their self-proclaimed, elite mobile phone services to the public. However, the services offered by most of these companies were way past poor. The coverage was not reliable and the service the companies offered were not compatible with most of the mobile phones available to the time.

By the early 1950s, mobile phones were restricted for use only in automobiles. The phones only worked in metropolitan areas and major cities. This was because major cities were equipped with large antennas that were located in the most central part of the city, and these large antennas were assigned specifically to these automobile phones. Automobile phones could only transmit a signal that spanned about 30 to 50 miles.

In 1983, when the Motorola DynaTAC 8000x arrived on the market, things changed. Developed and invented by Martin Cooper, this phone became the first mobile phone to be sold commercially and was priced at a whopping $4,000, which would be about $9,500 at present day. The phone weighed about 2 ½ pounds, and the dimension of the phone was 13 x 1.75 x 3.5. Although fairly large and expensive, the Motorola DynaTAC 8000x was considered small enough to carry around, thus making it the first mobile phone. After its debut, there has been a steady progression of the mobile phone, radio technology, and wireless technology to transform mobile communication into something that is very different from what it originally started from.

Because of the many technological contributions that have been invested in the advancement of the wireless communication and technology, the cell phone has become a hand-held system that can do virtually anything. From sending text messages, to video chatting, to surfing the internet, or finding the nearest shopping mall. The mobile phone has made life a lot easier with its unmatchable resources and constantly improving capabilities.

From Radio Phone to Cell Phone

Originally, the mobile phone began as a simple radio that transmitted signals with central antennas located in a specific range. This was not very effective, because the system only allowed extremely short-range calls, and coverage for these phones did not span very wide. Also, the maximum number of people that could talk on their phones at once did not exceed 50. Fortunately, a better alternative for mobile connectivity was introduced: Cell Technology.

Cell technology is a very unique, but seemingly simple form of technology that was developed to make wireless communication more accessible through the mobile phone. In contrast to the earlier telephony services, cell technology allows millions of people to use their cell phones simultaneously. The implementation of this innovative technology converted the mobile phone into a cell phone.

The cell phone is considered to be a full-duplex device. This means that the phone uses two separate frequencies to carry-out a phone conversation. One frequency is used for talking and the other is used for listening. This allows the caller and the recipient of the call to talk at same time, and even listen at the same time.

The process of cellular connectivity requires each city to be divided into small hexagon-shaped sections or "cells." The cells measure up to about 10 square miles, and are equipped with a tower, base station, and a building made to house the radio equipment. Cell phones operate within cells, and they can switch cells as they move around. The frequencies used by the cell phones are recycled and reused within the confines of the city. This is possible because the frequency will transfer over into the most adjoining cells, but not into cells that are more than one cell away. So, there is always coverage available and calls will rarely be out of range.

Today, cell technology has taken over the mobile phone market, and has boosted the call capabilities of mobile phones and has made connected calls a task of ease.

From Cell Phone to Smartphone

Smartphones are slowly replacing their predecessors. And it's due to their extremely advanced capabilities. It's no surprise that the progressions and popularity of the smartphone is constantly on the rise. These high-tech phones are able to do just about everything a computer can do. They have the ability to handle heavy weight computer applications like Microsoft office. Smartphone are able to sync and manage multiple email accounts, handle large files, and execute wireless capabilities and fast data speeds. These phones can do so much that you almost have to reconsider your personal computer or laptop.

Smartphones are only able to accomplish this because of their advanced operating systems. These operating systems grant you access to an endless world of information and accessibility that conventional mobile phones don't have the ability to do.

Early Years of the Smartphone

Device that integrated computing and telephony capabilities were first developed by Theodore Paraskevakas in 1971. He is the innovator of applying data processing, visual display screens, and concepts of intelligence to telephones, which marked the beginning of the final stages of the evolution of the smartphone. Paraskevakas's first low-tech exhibition, which involved a transmitter and a receiver to displayed various way to communicate through remote equipment.

The earliest device to demonstrate Personal Digital Assistant (PDA) abilities was the IMB Simon. This was only a prototype, but it functioned well enough to complete a very successful demonstration in front of hundreds at the COMDEX trade show that took place in 1992. Two years later, Bell South developed and released a similar model to the public, by the name of Simon Personal Communicator. PDA included faxing abilities capabilities, telephony capabilities, and applications like the address book, scheduler, calendar, calculator, clock, and LED touchscreen. This was considered to be the first smartphone in history. However, the term didn't become official until

1995.

One year later, the Nokia 9000 Communicator was released. This PDA device contained an operating system that was based on the GEOs V3.0. This operating system was originally used on the Nokia 2110. When the phone web browsing and the QWERTY keyboard were integrated, this became the blueprint for the first mobile device to be marked as a smartphone which was Ericsson R389. It was released in 2002.

The Dividing Factor

There are two small but major developments that sparked the evolution to the conventional mobile phone to the smartphone. The introduction of text messaging, multimedia messaging, and advance mobile operation systems created the grounds for separation between the regular radio/cellular phones to the smartphone.

The Development of Text Messages and Multimedia Messaging

Text messaging or Short Messaging Service (SMS) has definitely developed into something that has become very popular. Most people would rather send a text message before even using their phone to place a phone call. However, there was a time when the only text message that people could fathom, was sent and received through the mail. That all changed in 1933 when RCA Communications of New York sent the earliest documented text message to London through Radiotelegraphy. Within the first year, there were a total of 7 million words transmitted.

German engineer, Friedhelm Hillebrand developed the original concept of text messaging in 1985. His experiments led him to discover the amount of characters needed to successfully communicate through non-voice services. He discovered this by typing out random sentences on his typewriter, and then counting each character. After doing this numerous times, he found that each group of sentences averaged out to 160 characters. This quantity of characters became the limit that would be allowed for communication through non-voice, wireless communication.

In 1992, text messaging was first used by Neil Papworth, an engineer for the Sema Group located in the United Kingdom. He was able to send, "Merry Christmas" to a colleague's phone by way of Vadaphone through his personal computer. This inspired Radiolinja to develop the first SMS phone service to be offered to the public in 1994. The US picked up on this technology in 1995, and this led to Al Gore launching America's first major SMS network, Sprint Telecommunication Venture. Only a year later, international text messaging became available. Text messaging did not become a primary means of communication until 1998. Presently, text messaging is used by 74% of consumers worldwide.

Multimedia messaging is phone service that is utilized to send pictures, videos, music, ringtones, and other multimedia content. The first MMS was initially developed in Japan, and offered to the public in 2001. The first picture messaging service was offered through J-Phone.

Originally, Multimedia Messaging was developed to provide phone carriers with they call *captive technology*. This means that phone carriers would be able to charge consumers a fee each time they snapped a photo with their camera phone. The first deployments of this service were not successful, so was multimedia messaging service became available to the public. However, the initial deployments of multimedia messaging were plagued by technical problems and extremely unsatisfied consumers. The customers were experiencing problems with the quality of the multimedia messages they sent and received. A large number of consumers were complaining of being billed for sending messages that the recipient never received. Consumers also protested that they were billed for messages that had incorrect format and missing content. One of the main sources of these issues was the lack of MMS capable mobile devices during the early development of the concept. This was due to the underdeveloped technology that this phone feature had. However, in the recent years the MMS has been developed and improved, with the help some of the most elite technology companies in the world.

Multimedia messaging involves using the same technology that is used to send Emails. Whenever you send a MMS, your device will encode the multimedia content in the same format as an MIME Email. Then the message is forwarded to the phone carrier's service center and forward server, or Multimedia Messaging Service Center (MMSC). The phone carrier then sends the message to the MMSC of the recipient. From here, the MMSC will determine if the device receiving the message is enabled to for multimedia messaging. If the content will be accepted by the device, the media from the message is extricated and sent to an HTTP-front end, storage server. A text message containing the content's URL is sent to the handset of the recipient to prompt the device's WAP browser to open and extract the content from the embedded URL. If the receiver's handset is not MMS capable, the message will be forwarded to a web based service. Here the intended recipient will be able to view the content from the message through an internet service offered by their phone carrier.

At present, multimedia messaging has become just as popular as SMS. It is also one of the most used features on the mobile phone, followed by text messaging. Since the development of the MMS, there has been a massive increase in the spread of information, its compatibility with social media websites, email platforms, and other internet-based media sites will allow the MMS technology to flourish.

Mobile Operating Systems

Mobile Operating Systems or OS are programs that are assigned to the modern-day mobile devices. Operating systems are usually running in desktop computers, laptops, smartphones, tablets, PDAs, portable media devices, and the list goes on. The ingenious technology of mobile operating systems fuses together the capabilities of a regular desktop computer with the capabilities of mobile devices. This merge of capabilities includes but is not limited to, wireless internet, mobile navigation, GPS tracking, Bluetooth, voice recognition, media player, touchscreen capabilities, video recording, camera, calendar, and Email. The Read Only Memory (ROM) and Flash Memory chips found inside the circuit board or brain of the phone, supplies storage for the phone's operating system and features.

Most smartphones are comprised of two mobile operating systems. The primary user-facing software platform is supported by a secondary real-time operating system. This system operates the radio and other hardware. The first mobile operating system used dates back to 1973. The system was embedded in mobile phones to control the operations of the phone.

Android Operating System – Based on the Linux kernel and developed by Google, the Android operating system has one of the largest installed base on mobile devices worldwide. Most of the applications offered through the Android operating system are free and open-source software. Designed for smartphones, tablets, and other advanced mobile devices, the OS uses touch inputs that correlates with real-world actions. However, most of the Android devices are exclusive to only Android applications.

Apple Operating System or iOS – The Apple Operating System or the iOS is the second largest installed base in the world behind the Android. However, this is the most profitable installed based in the world. Phones that have this operating system are very expensive. iOS is also proprietary and a closed source. The user interface of iOS is based on the idea of direct manipulation. The touchscreen capabilities allow multi-touch gestures. All mobile phones and devices compatible with the Apple operating system are manufactured by Apple.

BlackBerry – The BlackBerry operating system is based on the QNX operating system. The system is closed source and proprietary. Unlike the Android OS, the Blackberry software is limited to the BlackBerry devices. All mobile phones and devices compatible with the Blackberry operating system are manufactured by Blackberry. The BlackBerry brand is mostly used by government officials.

Firefox – The Firefox operating system is an open source and is powered by Mozilla, a non-profit organization that is best known for its Firefox web browser. This is a Linux kernel based operating system exclusive to only smartphones, smart TVs, and tablet computers. The software is designed to supply consumers with an alternative community-based system compatible for mobile devices. This is accomplished by the utilization of open standards and various methods like JavaScript and HTML5 applications.

MIUI – The MIUI operating system is based on Google Android Open Source project. This operating system has a partial closed source. Developed by Xiaomi Tech, the MIUI operating system is installed of the Xiaomi smartphone, but can be found on a small selection of Android devices.

Sailfish OS – The Sailfish OS operating system is based on open source

Android libraries. It combines the Linux kernel and proprietary software that has been written by Jolla a smartphone manufacturer. This operating system is exclusive to all Jolla mobile devices.

Tizen – Tizen is sourced by the Linux Foundation and received support from the Tizen Association. Tizen is an operating system based on the Linux kernel that is made for mobile devices, smart TVs, and car entertainment systems. Tizen is an open source system that aims to provide the ultimate user experience for all of its associated devices.

Windows Operating System – Windows operating system is developed by Microsoft. It is closed source and proprietary. It has third largest installed base on smartphones behind Android and iOS. However, Microsoft has a number of operating systems for various smartphones. This phone includes computer based Microsoft applications such as OneNote, MS Excel, MS Word, and Outlook.

The Basics of a Cell Phone: Overviews and How-To's

*** this chapter covers a lot of basic capabilities of a cellphone; if
you are familiar with all basic functions of a cellphone, please skip
to the next chapter ***

Shopping for a cell phone can be a daunting task, but before you begin
your search for the perfect phone, it is vital that you are familiar with
the basic capabilities the current mobile phone. By acquainting
yourself with the essential software and capabilities of the modern
mobile phone, you will be able to make an educated decision in picking
out the phone right for you. A plus is that this may help you save
money in the long-term.

Overview of the General Phone Applications

A mobile application is a computer program designed to work on
mobile devices. Mobile phones are run by various applications.
Subsequently, modern smartphones are can be considered computers
more than they can be considered mobile phones. In a conventional
sense, it is only right that smartphones contain a fully-functional OS of
their own. Mobile operating systems make the advanced features of
mobile phone available through mobile applications.

When making phone applications, developers have to be conscious of the capabilities of that phone and the inabilities of the phone. Developers must be aware to the variety of phones that are out there on the market. The frequent changes in the mobile operating systems can make it difficult for some developers to keep up. Mobile application development requires use of specific amalgamated development environments. Mobile apps are first tested within the development environment using emulators. The use of emulators offers developers an inexpensive way to test applications on phones that they have no access to.

Developers use mobile user interfaces to help keep the user as the focus of the development and design. The mobile user interfaces grants the developers the ability to mimic how the user's manipulation will affect the developing application. This helps developers give users a user-friendly mobile application.

Some of the first mobile applications appeared on the Personal Digital Assistant (PDA) or handheld PC in 1984. It wasn't until 1994 that PDAs gained the ability to place phone calls. This was the earliest form of the smartphone, because during that time nearly all PDAs had the ability to do what the smartphone of today can do. PDAs offered applications like touchscreen capabilities, a QWERTY virtual keyboard, handwriting recognition, stroke recognition.

Today, the mobile apps development landscape and the mobile app marketplace has sky rocketed. Thanks to the many contributions of mobile operating system developers like Google's Android OS, Apple's iOS platform, Microsoft's Windows Phone, and hundreds of smaller third-party companies and teams that sell mobile applications through the major mobile application markets. Most of the phones on the market today are programmed with some essential applications that are universal to all mobile devices. Let's take a look at some of the most used applications below.

Calculator

The calculator on your phone works just like a generic pocket calculator. The calculator is a free application already programmed into your phone. Even some of the more basic mobile phones contain this application. You can locate the calculator application one your phone's home screen or the phone's application menu.

This phone has the basic capabilities to carry out math problems such as addition, subtraction, multiplication, and division. Some calculator apps will be able to handle more advanced math problems like square roots and radicals. However, this depends on the operation system.

If would like to have a calculator that is a little more advanced, then there are plenty of options for you located in your phone's application store.

Calendar

The calculator is a free application already programmed into your phone. Even some of the more basic mobile phones contain this application. The calendar application can be located in the application menu of your phone. Just simply touch the calendar icon to schedule events, set reminders for important dates, and much more.

Depending on the operating system, you may be able to sync your calendar on your phone with your calendars located on your computers and other mobile devices. Operating systems such as Android and Windows are great examples of the syncing feature. Be sure to check out your phone manual for a better understanding of how the features of the calendar application work.

Cloud Computing Services

Cloud computing services are becoming very popular and have evolved drastically since the 1990s. Your phone's operating system usually has its own cloud service. You will have your own cloud service once you phone is activated. It will be secured by your username/email address and a password. You will be able to upload files, media, contacts, mobile applications, and other important information to the cloud service. This is convenient to users because you will be able to access everything saved to the cloud from any mobile device or computing system. Also if you were to switch phones you will be able to download everything from the cloud service to your new phone with no data loss.

Contacts

Each contact that you add to your contact list will be automatically synced with the Voice-To-Call and Voice-To-Text features on your phone. All you have to do is speak the contact's name into the microphone of the mobile device along with the intended command, and it will begin completing the command. For example, if I wanted to call my brother John, I would pull up the voice command application. Then, all I would have to do is speak, "Call John," into the phone. The phone will begin placing the call.

Global Positioning System (GPS)

This is a satellite based feature that is available to all mobile phones. This system is comprised of a network of approximately 30 satellites that have been placed into orbit by the U.S. Department of Defense. This feature provides clock synchronization which allows the time on your phone to be automatically updated. This is critical for synchronizing its spreading codes with other base stations to enable inter-cell handoff and provision for hybrid GPS/cellular position recognition that is necessary for emergency calls and other applications.

Lock Screen

The lock screen is a security feature of the user interface implemented to mostly all smartphone. Its purpose is meant to provide security to the personal information in your phone. When this feature is activated, it will require a password to be entered whenever the phone is used after the screen times-out or whenever the phone is powered back on. This feature can be activated under the 'SETTINGS' options.

Media Player

The media player is an application that plays the music, videos, and podcasts downloaded to the phone. You can also receive live stream support through the later media players. Most media players are able to automatically detect the media stored in your phone and on your SD card, and add the media to the library. This application is controlled by commands such as play, pause, skip forwards, skip backwards, shuffle, repeat, and loop.

Navigation

The GPS navigation application is responsible for precisely calculating the geographical locations through GPS satellites. This applications gives you access to step-by-step directions to and from any location of your choice. You are also able to access traffic congestion maps, street maps, and directions via text or speech.

Photo Gallery

The photo gallery is the application where you can view the pictures that have been downloaded, Bluetoothed, or saved to your phone from the camera. Just like the media player, the photo gallery is able to automatically detect the media stored in your phone and on your SD card, and add the photo gallery.

Marketplace/Store

This is the digital distribution platform for mobile applications. Each smartphone has this feature. Mobile Marketplaces mimic the layout of an online store, where users can browse through these various app categories, view details about each mobile application, (e.g. number of downloads, ratings, reviews, etc.), and access applications (e.g. ringtones, games, and any applications that you think would be useful to you and your phone). The applications are offered as an automatic download, and the installation of the application begins. Some marketplaces may also allow the operating system to spontaneously remove an installed application from device the mobile device under specific conditions, to prevent malicious software.

Web Browser

A mobile browser is a web browser designed for use on a mobile device such as a mobile phone or PDA. Mobile browsers are augmented to present web content that is suitable for small screens of mobile devices.

Overview of the Codes Assigned to Your Phone

Every cell phone that you come in contact with has unique, alphanumeric codes that are only assigned that individual phone. These codes are utilized to identify the phone, phone carrier, and owner. Familiarize yourself with these codes below:

System Identification Code (SID)

This is a unique 5-digit number that is allocated to each phone carrier by the Federal Communications Commission (FCC). This code is programmed into the phone at the time of purchase. This unique code is used to identify cellular networks in identifiable areas. System Identification Codes are transmitted by several base stations in order to locate the cellular networks. When the phone is powered on it listens for a signal, and once the signal is identified and received, the phone investigates the SID. It compares the SID signal to the SID that is recorded in the phone.

Electric Serial Number (ESN)

This is a unique 32-digit number that is allocated to the phone by the manufacturer. The number is programmed in the phone at the time of its creation.

Mobile Identification Number (MIN)

This is a 10-digit serial number that is allocated to the phone, once a phone number is assigned to the phone by the phone carrier. This number is used to identify phones that are operating in cooperation of TIA (Telecommunications Industry Association) standards and PCS (Personal Communications Service) technologies.

International Mobile Station Equipment Identity (IMEI)

An IMEI is your 14 to 16 digit serial number that will be unique to your mobile device. The use of this number has been implemented to help eradicate the use of stolen mobile devices. This is a unique number used to identify 3GPP, iDEN phones, and satellite phones. This number is normally located inside the battery compartment of the mobile phone. It can also be displayed on the phone screen by entering in '#06#'. You can also find this number by going to the settings menu and looking the phone's operating system.

Inside the Phone

The basic components that will bring your mobile phone to life as follows:

Random-Access Memory (RAM)

The memory in your phone is where the software and programmed into your phone and the data runs. Random-Access Memory is a type of computer storage data that houses the operating system software and the applications on your phone. RAM allows the data in your device to be interpreted at the same rate simultaneously no matter the order the data is accessed. This memory runs at a very fast pace. Mostly all mobile and stationary computer devices have RAM. An increased amount of RAM allows for more complex software to be run on the device.

Antenna

Embedded in the handset case, the antenna is a very important technology component to the mobile phone. The antenna provides the phone with all of its wireless capabilities. The antenna is the base of all wireless communication, and enables the phone to perform at the best quality.

In a wireless system, an antenna converts guided radio wave energy (such as a signal traveling in a coaxial cable for television) to energy that is emitted or "radiated" out into free space. An antenna also does the reverse—it receives radio waves from the air and feeds them into the devices that detect, decode, and amplify them. In a cell phone system, there is one antenna in the handset and another in the base station tower. Both of these antennas transmit and receive waves.

Keyboard/Keypad

The QWERTY virtual keyboard, the QWERTY the keypad, or the keypad is available is the input device on your mobile phone. Here you will be able to communicate message, type in phone numbers, and input other commands.

Microphone

The microphone is the part of the phone that you will speak into. This will be where you will talk on the phone, and send out voice commands.

Liquid Crystal Display (LCD) Screen

The LCD screen will be where you will be able to view all of the actions, commands, and applications to your phone. This is what makes the phone so attractive. However, this is one of the most fragile features of the phone. It is very susceptible to damage, so it is very important to purchase a reliable and sturdy phone case to protect your phone and screen from damage.

Battery

The battery provides electricity and power to your phone. You can monitor your battery life through the icon that will be located at the top left corner of your phone's display screen. Your phone battery will likely be composed of lithium.

Speaker

The speaker is where you will be able to listen to phones calls, music, and hear notifications.

Circuit Board

The circuit board is the brains of the phone. It consists of several chips that translate audio signals.

How to Work the Most Important Applications on Your Phone

These are general instructions. Specific procedures will differ by operating system. Check your phone manual for the instructions unique to your phone

Alarm/Clock

The alarm clock is a basic feature on the phone, and a free application already programmed into your phone. Even some of the more basic mobile phones contain this application. Here you will be able to set the time, and set alarms. You can also see the time on your Home screens by adding a Clock widget.

To set new alarm:

- Open your phone's Alarm/Clock application.

- At the bottom of the screen, tap the add icon that is usually indicated by a "+" sign.

- Enter in the time that you would like for the alarm to go off.

- Next, you will need to set the alarm to repeat. Depending on the OS there will be a dropdown menu titled, 'REPEAT' with a list of the days of the week. Select the days that you will need for this alarm to go off.

- Then you can choose the ringtone that you would like for the alarm. This option is available under the 'SOUND' menu.

- You will then need to label the alarm. Touch the text box located under the 'LABEL' or 'NAME' title. Type in the name you prefer.

- Set the time you want, and then select 'OK,' 'SAVE,' or 'DONE'.

- Once the alarm is saved, you will be able to view the alarm in a list. From here switch the alarm from 'OFF' to 'ON'.

Dialer/Phone

All mobile phones have the capability of placing and receiving phone calls. This is the most essential feature of any phone. The typical phone call is placed by picking up the cell phone and dialing in the recipient's phone number. This can be done through the touchscreen QWERTY keyboard, or the alphanumeric touchpad. More advanced mobile phones offer the feature of voice dialing. You can place a call through the phone by simply speaking the recipients' name (recipient's name and number must be saved in your contacts) or speaking each digit of the phone number into the microphone.

You will be able to place a phone call whenever there is a paid cell phone subscription assigned to your phone. If there is an interruption is your service, bad reception, the phone is powered off, or if you are out of range of your phone carriers' coverage areas, you will not be able to place a call.

Usually, you will not be able to place a call while the cell phone is busy running another application. However, this rests solely on the capabilities of the phone. Since some phones do allow phone calls to connect while the phone is busy running applications.

The easiest way to place a phone call is to begin at the home screen. Select the 'DIALER' or the application that is used to make phone calls. From here, you will need to make sure that you know the full phone number of the person you are calling.

There are three types of phones calls that can be made from your phone, local, long distance, and international. To determine if the call you are placing is local, long-distance, or international, you will need to check the country code, area code, and/or city code.

To get a better understanding of what these phone numbers will look like, check out the examples below:

To place a LOCAL Phone Call

To determine if your phone will be connected locally, be sure that you have an area code associated with the phone number you wish to call. If there is an area code, check to see if the area code matches with yours or it the area code is out of state. If the area codes are the same, then this will be considered a local phone call. However, the call may still be considered local if there is a different area code. Most cities have boroughs or suburbs that are located no less than 10 minutes from major metropolitan areas. These suburbs/boroughs will usually have area codes that differ from the main city. If this is the case, this will still be considered a local call.

To place a local call, you may enter either the last seven digits of the phone number, or you can enter the full ten digits. For example, if I live in Sacramento, CA *(area code 916),* and I want to call my grandmother that lives two blocks away, I will need to enter in the phone number as follows:

To call phone number **(916) 555-555** or **555-5555:** *I will first, dial area code '***916***'. Next, I will dial phone number, '***555-5555***'. Finally, I will select 'SEND".*

To Place a LONG-DISTANCE or DOMESTIC Phone Call

If the call that you wish to place is long-distance, out of state or in another city, you will need to enter the full phone number or all ten digits. For example, if I live in Sacramento, CA *(area code 916)*, and I wanted to call my cousin in Memphis, TN *(area code 901)*, I will need to enter the phone number as follows:

To call phone number **(901) 555-5555:** *I will first, dial area code '***901***'. Next, I will dial phone number, '***555-5555***'. Finally, I will select*

'SEND'.

To Place an INTERNATIONAL Phone Call

If I am placing an international call to a mobile phone, I will dial 011, and then the country code for the country I am calling, the area or city code, and the phone number. However, there is a general rule of thumb when calling international. If I am in America, and I am placing a call to a cell phone that is native to the Phi Phi Islands, Thailand, I will need to type the number in as follows:

To call phone number **(+66) 2-555-5555**:

*First, I will dial, '***011***'. Next, I will dial country code, '***66***.' Then, I will dial city code '***2***'. Then, I will dial phone number, '***555-5555***'. Finally, I will select 'SEND".*

International phone numbers have separate city codes for landlines and separate city codes for mobiles phones. For example, if I live in Lagos, NG and I wanted to call a local land line phone, the city code will be different from my mobile phone even though we live in the same town. Most city codes contain two or three digits. However, some international phone numbers do not have a city code. If you feel that you are not capable of sending a text message internationally, check your phone carrier's website for further instruction, or to find out if your phone carrier has special instructions for sending international calls.

***Poor call quality is a common issue with cell phones. Sometimes calls may have terrible reception, calls may be dropped, or calls be unsuccessfully connected. This problem usually has everything to do with your phone carrier. However, if you are experiencing frequent problems with your call quality, you will first need to attempt to solve the problem on your own.

To troubleshoot you will need to do the following:

If you are experiencing frequent poor call quality, check the signal strength of the phone. This can be done by observing the signal indicator (usually 4 to 5 bars ascending in size) that is usually located at the very top of the phone's display screen. If the problem persists or you need help with more options for troubleshooting, please refer to: Mobile Phone Signal to be advised.

Email

Setting up your email account on your mobile phone will be very easy if you have an email account associated with the operating system programmed to the phone. Start the email program on your phone by tapping the 'EMAIL' icon associated with the operating systems email provider (e.g. Gmail for Google's Android, Outlook for Microsoft's Window, iCloud for Apple's iOS, etc.). The phone will simply prompt you to enter in your login information for your email service provider.

Once you have entered is the login information, you should be able to access your email easily. However, if you do have other email accounts associated with other operating systems you can either download the email application to your phone's marketplace if it is available, or you can set it up through POP3 or IMAP.

Setting Up POP (Post Office Protocol) or IMAP (Internet Message Protocol)

You can set up email on a wide selection of phones with internet capabilities through POP or IMAP. Here are some universal instructions for setting up email on your mobile phone. Some setups slightly differ from one phone to another, but the general instructions are pretty much the same.

Start the email program on your phone by tapping the 'EMAIL' icon. The email application is usually located in the application menu of your phone. Choose the option to add an account. If you need help finding the email app, use the documentation that was included with your phone or tablet.

Use the following information to help you set up your email.

- **Email address**: Enter in your full email address when prompted (for example, tony@phoneemail.com).

- **User name**: Your user name and email address is the same thing. You may be asked for both or either one of the two (for example, tony@phoneemail.com).

- **Password**: This is the password for your email account.

- **Incoming mail server settings**: These are your incoming (POP3 or IMAP4) server settings.

Text Messaging and Multimedia Messaging

You will be able to send a text message whenever there is a paid cell phone subscription assigned to your phone. If there is an interruption is your service, bad reception, the phone is powered off, or if you are out of range of your phone carriers' coverage areas, you will not be able to send a text message. Usually, you will not be able to send a text message while your cell phone is busy running another application. However, this can depend on the capabilities of the phone.

The easiest and most common way to place a send a text message is to start at the home screen. Select the 'Messaging' icon or the application that is used to send text messages. From here, you will need to make sure that you know the phone number or email address that you are sending the text message to.

There are two types of messages that can be sent from your phone, to numbers that are local, long distance, and international or to any email address. You will be able to send a Multimedia Message Service (MMS) or a Short Message Service (SMS). If your area code is the same as the recipient's or if the recipient lives in the same city, this will be considered a local phone call. If the individual is located out of state or across the country, then this will be considered a long-distance phone call. International calls apply to phones calls that are placed to recipients out of the country.

Check your cell phone carrier to see how much text messaging and multimedia messaging costs, or if messaging is included in your phone plan. If your phone plan does not allow you to send text messages, each message you send or receive will cost about 10¢ to 20¢ per message. For international or roaming the rate may increase.

Sending a Text Message

All mobile phones have the capability to send and receive text messages. Text messaging is one of the essential features of the mobile phone. Phones will in regards to the menu options and buttons, but the procedure of sending a text message is pretty much the same. A text message can be sent by picking up the mobile phone and dialing in the recipient's phone number through the touchscreen QWERTY keyboard or the alphanumeric touchpad

The standard text message or SMS (Short Message Service) involves composing and sending concise, electronic messages that are transmitted between two or more mobile phones, or a computer system. Text messages are usually limited to 140 to 160 characters (numbers, letters, symbols, and spaces). If your text message exceeds the character limit, it is automatically separated into several text messages. Each text message is usually sent one after the other until the entire text message is sent to the recipient.

More advanced mobile phones offer the feature of 'Talk to Text'. You can send a text message by simply speaking the recipient's name (recipient must be saved in your contacts) or speaking each number of the phone number into the microphone. You can then relay the message through the microphone, and the phone will type it out to the best of its abilities. All you will have to do at this point is press or say send.

You will be able to send a text message whenever there is a paid mobile phone subscription assigned to your phone. If there is an interruption in your service, bad reception, the phone is powered off, or if you are out of range of your phone carriers' coverage areas, you will not be able to send a text message. In most cases, you will not be able to send a message while the mobile phone is busy running another application. However, if the phone is still in the process of sending the message, you will be able to run applications as normal.

The easiest and most common way to send a text message is to start at the home screen. Select the 'MESSAGING' application that is used to make send text messages.

To Send a LOCAL Text Message

To send a text message locally, be sure that you have an area code associated with the phone number you wish to call. If there is an area code, check to see if the area code matches with yours or it the area code is out of state. If the area codes are the same, then this will be considered a local phone call. However, the call may still be considered local if there is a different area code. Most cities have boroughs or suburbs that are located no less than 10 minutes from major metropolitan areas. These suburbs/boroughs will usually have area codes that differ from the main city. If this is the case, this will still be considered a local call.

To send a text message locally, you will need to enter the full ten digits of that phone number when prompted. Unlike placing a phone call locally, failing to include the area code will cause the text message to be unsuccessful. It is vital to that the entire phone number is entered in when prompted. For example, if I live in Vallejo, CA *(area code 707)*, and I want to send a text message to my sister that lives on the other side of town, I will need to enter in the phone number as follows:

To send a text message to phone number: **(707) 555-555 or 555-5555**: *I will first, dial area code '707'. Next, I will dial phone number, '555-5555'. Finally, I will type in the text message that I would like to transmit, and select 'SEND'.*

To Send a LONG-DISTANCE or DOMESTIC Text Message

To send a text message to a long-distance phone number, you will need to enter the full phone number or all ten digits. For example, if I live in Pasadena, CA *(area code 626)*, and I wanted to call my cousin in Olive Branch, MS *(area code 662)*, I will need to enter the phone number as follows:

To a send text message to phone number: **(662) 555-5555**; *Dial area code '662'. Then dial phone number '555-5555'. Finally, I will type in the text message that I would like to transmit, and select 'SEND'.*

To Send an INTERNATIONAL Text Message

If I am sending a text message to an international phone number, I will dial 011, and then the country code for the country, the city code, and the phone number. However, there is a general rule of thumb when calling international. If I am in America, and I am placing a call to a cell phone that is native to London, UK, I will need to follow type the number in as follows:

To a send text message to phone number: **(+44) 20-555-5555**: *First, I will dial, '011'. Next, I will dial country code, '44.' Then, I will dial city code '20'. Then, I will dial phone number, '555-5555'. Finally, I will type in the text message that I would like to transmit, and select 'SEND'.*

International phone numbers have separate city codes for landlines and separate city codes for mobiles phones. For example, if I live in Lagos, NG and I wanted to call a local land line phone, the city code will be different from my mobile phone even though we live in the same town. Most city codes contain two or three digits. However, some international phone numbers do not have a city code. If you feel that you are not capable of sending a text message internationally, check your phone carrier's website for further instruction, or to find out if your phone carrier has special instructions for sending international calls.

***Be sure to check the status of the text message after you select 'SEND'. Sometimes the text message will fail to send and will automatically save to the 'OUTBOX'. In most cases, the phone will be able to send the message on its own after some time has passed. However, if the message fails to send after an extended amount of time, you will first need to attempt to solve the problem on your own.*

To troubleshoot you will need to do the following:

Select 'OUTBOX' from the SMS menu. Locate the failed message and select to review. Once you have reviewed the message, select 'RESEND'.

If the text message fails to send again, check the signal strength of the phone. This can be done by observing the signal indicator (usually 4 to 5 bars ascending in size) that is usually located at the very top of the phone's display screen. If the problem persists or you need help with more options for troubleshooting, please refer to: Mobile Phone Signal to be advised.

Sending a Multimedia Message

All mobile phones have the capability to send and receive multimedia messages. MMS is one of the essential features of the mobile phone. Phones will in regards to the menu options and buttons, but the procedure of sending a multimedia message is pretty much the same. An MMS can be sent by picking up the mobile phone and dialing in the recipient's phone number through the touchscreen QWERTY keyboard or the alphanumeric touchpad, and attaching whatever media type that you would like to send to the message.

You will be able to send a multimedia message whenever there is a paid mobile phone subscription assigned to your phone. If there is an interruption in your service, bad reception, the phone is powered off, or if you are out of range of your phone carriers' coverage areas, you will not be able to send a text message. In most cases, you will not be able to send a message while the mobile phone is busy running another application. However, if the phone is still in the process of sending the message, you will be able to run applications as normal.

The easiest and most common way to send a multimedia message is to start at the home screen. Select the 'MESSAGING' application that is used to make send messages.

To Send a LOCAL Multimedia Message

To send a multimedia message locally, be sure that you have an area code associated with the phone number you wish to call. If there is an area code, check to see if the area code matches with yours or it the area code is out of state. If the area codes are the same, then this will be considered a local phone call. However, the call may still be considered local if there is a different area code. Most cities have boroughs or suburbs that are located no less than 10 minutes from major metropolitan areas. These suburbs/boroughs will usually have area codes that differ from the main city. If this is the case, this will still be considered a local call.

To send a multimedia message locally, you will need to enter the full ten digits of that phone number when prompted. Unlike placing a phone call locally, failing to include the area code will cause the multimedia message to be unsuccessful. It is vital to that the entire phone number is entered in when prompted. For example, if I live in New York City, NY *(area code 212)*, and I want to send a picture to my sister that lives on the other side of town, I will need to enter in the phone number as follows:

To send a multimedia message to phone number: **(212) 555-555** or **555-5555**: *I will first, dial area code '212'. Next, I will dial phone number, '555-5555'. Finally, I will attach the picture or the media that I would like to transmit, and select 'SEND'.*

To Send a LONG-DISTANCE or DOMESTIC Multimedia Message

To send a multimedia message to a long-distance phone number, you will need to enter the full phone number or all ten digits. For example, if I live in Jackson, TN *(area code 731)*, and I wanted to call my cousin in Orlando, FL *(area code 407)*, I will need to enter the phone number as follows:

To a send text message to phone number: **(407) 555-5555**; *Dial area code '407'. Then dial phone number '555-5555'. Finally, I will attach the picture or the media that I would like to transmit, and select 'SEND'.*

To Send an INTERNATIONAL Multimedia Message

If I am sending a text message to an international phone number, I will dial 011, and then the country code for the country, the city code, and the phone number. However, there is a general rule of thumb when calling international. If I am in America, and I am placing a call to a cell phone that is native to the Hong Kong, CHN, I will need to follow type the number in as follows:

To a send text message to phone number: **(+852) 5555-5555**: *First, I will dial, '***011**'. *Next, I will dial country code, '***852.***' Then, I will dial phone number, '***5555-5555**'. *Finally, I will attach the picture or the media that I would like to transmit, and select 'SEND'.*

International phone numbers have separate city codes for landlines and separate city codes for mobiles phones. For example, if I live in Lagos, NG and I wanted to call a local land line phone, the city code will be different from my mobile phone even though we live in the same town. Most city codes contain two or three digits. However, some international phone numbers do not have a city code. If you feel that you are not capable of sending a text message internationally, check your phone carrier's website for further instruction, or to find out if your phone carrier has special instructions for sending international calls.

***Be sure to check the status of the text message after you select 'SEND'. Sometimes the text message will fail to send and will automatically save to the 'OUTBOX'. In most cases, the phone will be able to send the message on its own after some time has passed. However, if the message fails to send after an extended amount of time, you will first need to attempt to solve the problem on your own.

To troubleshoot you will need to do the following:

Select 'OUTBOX' from the SMS menu. Locate the failed message and select to review. Once you have reviewed the message, select 'RESEND'.

If the text message fails to send again, check the signal strength of the phone. This can be done by observing the signal indicator (usually 4 to 5 bars ascending in size) that is usually located at the very top of the phone's display screen. If the problem persists or you need help with more options for troubleshooting, please refer to: _Mobile Phone Signal_ to be advised.

Business or Pleasure: What is the purpose of your cellphone? **Bonus** A Quick Budgeting Guide For Cellphone Buyers!

Whew! Now that is a lot of information to take in about the technicalities of the mobile phone. Now, it's time to come down from a technical level and settle to a more personal level. It's time to think about what mobile phone will work for you, and the mobile phone you choose will serve its intended purpose in your everyday life and business, activities, and responsibilities. This is the part that can be quite confusing and can cause so feelings of indecisiveness with the numerous options that are available.

Knowing what you want

Choosing a cell phone that is just right for you can end up being a very difficult. It's no surprise since the market is flooded with many options. Today, there are over 100 cell phone manufacturers and thousands of phone carriers. With this vast selection, it can be easily understood why picking out a cell phone can be a daunting task. Mobile phones are offering accessible features unlike any device on the market today.

Consumers now have easier SMS and MMS services, web browsing, GPS navigation, and social networking while staying in tact with work and their everyday lives. The smartphones are leading the charge due to operating systems like Windows, Android, and Apple. These phones' advanced operating systems can run many varieties of applications that offer you access to information that are not easily accessible to you.

These phones have helped people become more successful with business, time management, money management, and other important factors that human beings depend on for sustenance. However, conventional or lower-tech phones are not far behind. They offer convenience with their large screens, QWERTY keyboards, and web browsing capabilities. They also contain email applications, but they are not as robust as the email applications that you will find accessible with the smartphone. However, conventional phones offer email applications that are a lot simpler to use.

Hold off on dashing out and buying that new phone, there is still a lot of information to cover before you get to the fun part. It is important to consider a few underlying factors when your decided to purchase a mobile device.

When you're ready to buy a phone, you'll first have to decide which of the two types, conventional mobile phone or smartphone. Choose a conventional model if you mainly need voice and text-messaging capability, and perhaps a music player and camera. Smart phones, with their advanced operating systems, larger displays, QWERTY keyboards, and other computer-like features, are a better choice for people who need frequent access to multiple e-mail accounts, a sophisticated organizer for appointments and contacts, the ability to open Office documents, and Internet-based services. One compelling advantage of most smart phones is their ability to access a host of applications consisting of productivity tools, shopping, multimedia, games, travel, news, weather, social, finance, references, etc.

What do you use your phone for?

When picking out a phone it is very important that you know how it will fit into your lifestyle. Knowing what you will use your phone for, how you will use your phone, and how often will prepare you for the road ahead. If spending money is a science, and planning for something that you would like to spend that money on is Quantum Physics mixed with a serving of Calculus on the side. Thinking about all the things that you may or may not do with your phone will require you to sit down and really think about who you are. Will you use your phone for leisure, for business, or for the basic purpose? These are some of the questions that you will need to ask yourself before picking up that credit/debit card.

If you text more than you talk, why would you buy get a phone plan that's unlimited talk and text and vice versa? If no one ever calls you and you rarely place calls then why wouldn't you get a mobile plan that's suitable for your use? The scariest thing about a majority of consumers today is that most shopping is done on impulse anyway. There is no thought behind their purchases, they just simply buy.

Blindly picking out your phone will be just like walking into a grocery store hungry. You will end up finding a lot of stuff that you don't really need. So, sit down and write out the things that you do every day and see how your phone can accommodate you.

You will need to also look at other features like the screen size, applications, and other features of the phone. Ask yourself if this phone will be the right one for you. Ask self if this phone will phone be a distraction, and aide, or a helpful business tool. Are you going through a life changing event such as working from home or retirement? If so, you may need to go for a phone that compliments that transition. If your lifestyle is relaxed you need to have a phone and a phone plan that is relaxed as well.

Budget

Your budget is affected by a few factors, your shopping style, your preferred phone service, and your needs. You will consciously or subconsciously base your purchasing decision on these three factors. However, to make the budget work the best for you it is vital to be able to channel these factors in the direction that you need to.

There are new and enticing phones being released frequently that offer greater accessibility than previous or outdated models. Even the outdated models contain some appealing features. But to pick the perfect phone, you will need to sort through the available selections, sort through all of your wants and needs, and come to a sound, mature, and responsible decision.

This may seem like a hard task, and many will not know where to start. However, the easiest place to start is to know your budget, the factors that will affect your budget, and what types of strategies to adapt to make your budget work for you. Let's become knowledgeable of the things that will come into play so that you can strategize your purchasing decision. Here are a few basic tips of budgeting that you can adopt and apply to your phone shopping.

Write Down All Expenses

Bills are the heart of your budget, and what keeps us motivated to keep making money. Whether it is a food bill, rent, car note, gas, education, citations, credit card payment, it doesn't matter how big or small the bill is. Bill should have a permanent place in your budget. Adding your bills and expense to your budget will also keep you up-to-date on the due dates as well. Many people are late on bills simply because they forgot about the due date. It not that they weren't can't afford to pay the bill, it's just that there are other things that they would rather spend their money on. To get a good grasp on your finances, include everything that deals with your money in your budget.

It would also help if you categorized all of your bills based on how you spend and life events. For example, birthdays and other special occasions can be something that you set aside or budget. Or maybe you can pencil in a little money to include for that perfect phone that you've been looking to buy. To make the process a lot faster and easy on the eyes, you can enter your budget into a spreadsheet. This will help with organizing the categories of each expense as well, and make looking at your budget less stressful.

Be Consistent

Never stop budgeting. This is the most important thing to remember about budgeting. The rule is to never spend more than you earn. However, many tend to forget this rule even though it's pretty straightforward. Spending more than you earn breaks the cardinal rule of money management, because that puts you in the category of deficit spending. It doesn't matter if you are a middle class citizen or a millionaire, if you are always spending more than you ear, you will always be in debt. Another helpful tip is to add up all of your expenses and compare them to how much money you actually make to determine if your bills are trumping your earnings. If this is the case, then you must reconstruct your budget. You will need to make the must-have your top priority, along with cutting off all of your luxuries and things that you can live without. There should be no shortcuts taken, or you will only be cheating yourself.

Have a Nice Treat

All budgeters know that treating yourself will help the budgeting process go a lot smoother than if there was no treat at all. However, these privileges are planned and added to the budget. They are not just based on an impulsive purchase. Treating yourself within a budget will help relieve the stress of paying bills and sticking to your budget. The best ways to think of these rewards are as treats for outstanding money management. So, when you are making your budget think of something that you would really want that can safely fit into your budget without jeopardizing any of the money for the necessities. Choose the reward carefully and apply it to something constructive, and something that really makes you happy. It could be a hobby, shoes, a dance class, or anything that makes you tick. Just make sure it's budget friendly and modest. So, don't deprive

Plan for the Worst

Those individuals that are keen on budgeting know that planning for the worst is vital. Life brings along many unexpected events and accidents. They are always lurking somewhere waiting to attack. It is the best practice to plan for those unexpected events accordingly, and neutralize the effects of those events that you were never looking forward to. You can do this by maintaining your greatest assets frequently and to the best of your abilities.

Avoid and Ignore the Delusions

To become a master at budgeting, you will have to change your way of thinking and perception of budgeting altogether. What is the first thing that comes to mind when it comes to budgeting for you? Do you secretly hate the idea because you wish you had more money to spend? Are you afraid that creating a budget and sticking to it will change your comfortable lifestyle? Well, it's time to get rid of those beliefs, and realize that even the richest people in the world budget. Just know that budgets are not as constricting as you think. You can still have fun on a budget, and a carefully thought-out and planned budget will allow you some special rewards in the end. These rewards are your aim. However, it is best to think of your budget as a spending plan unique to you. With this plan you will be able to accumulate the great skill of money management. This will give you a more responsible way of spending money. Having a set budget will cut out impulsive spending, and give you control over your money, expenses, financial goals, eliminate debt, and help all of those bills fade away.

What type of Shopper Are You?

When shopping for a mobile phone it is important to keep three things in mind your budget, your shopping strategy, and your needs. Your budget and needs are the most important of the three. This is important to keep in mind because it is almost inevitable that the phone you decide to invest in will be something that you would want to stick with for a while. So, the best place to start is to know who you are as a shopper, and what type of shopping strategy to adapt to pick the best phone possible.

Let's take a look at the various types of shoppers to determine your purchasing style. Knowing the type of shopper you are will help you decide your strategy when you finally go decide to go out and purchase your phone. This will also supplement your budget and help you determine where you should buy your phone.

- **Discount Shopper** – This shopper is the type of shopper that loves to saves where there is any opportunity available. This person loves to make every purchase budget-friendly. Discount shoppers will always have coupons, discount codes, the latest information of sales, and all price comparison information handy at all times. Hunting down deals is second nature to this individual, and that's a good thing because this person can see all opportunities to save. It's best to adapt to this style if your budget is really low but you want to buy a phone that is much more expensive than you can afford. Since the mobile phone market is so full of options, it won't be hard to find the most budget friendly phone with the most advanced features.

  Best Places to Shop for a Phone: Anywhere and/or Bidding sites (e.g. EBay, Quibids.com, Craigslist, Propertyroom.com, Ubid.com, CowBoom.com, and Listia.com); Pre-paid phone service providers; Stores with marketplaces (e.g. Amazon.com, BestBuy, etc.)

- **The Researcher** – This person partakes in long and extensive research that can last for months. They are keen to all off the latest and future product releases. They carefully and articulately strategize when it comes to shopping and purchase only when they feel the time is right. They might make a decision to buy something right away or they make wait for months to purchase the thing that they've had their eye for a while. This is one of the smarter and less stressful ways to shop.

Best Places to Shop for a Phone: Anywhere

- **The Impulsive Buyer** – If this shopper sees something they buy it. However, this shopper is not the best kind of shopper to be if you have a very tight budget. If you adapt this shopping strategy when purchasing a phone, you might end up with something that you will regret. So, if you buy on impulse now is the time to take a step back and think about a different shopping strategy that will work for your benefit. Another option is sitting down and thinking very hard about what you truly want in a phone. Write this information down and take it with when you visit your preferred phone service provider and ask a salesperson for their insight and help.

Best Places to Shop for a Phone: Mobile service providers or any place where there will be staff to help.

- **The Negotiator** – This shopper knows that no price is every set in stone. Everything is always debatable. Their main goal is to get what they want at the cheaper price by bargaining with the salesperson until they can't bargain anymore. If you think that you have what it takes to walk into any store and talk the salesperson down to a price reasonable for you then go for it! However, this may not work so it will be better to settle for purchasing a phone through bidding.

Best Places to Shop for a Phone: Anywhere and/or Bidding sites (e.g. eBay, Quibids.com, Craigslist, Propertyroom.com, Ubid.com, CowBoom.com, and Listia.com)

- **The Loyal Customer** – This customer buys everything from the same place. Not only that they have a slew of membership cards, discounts, and reward points. This shopper knows what stores suit their needs and they are going to stick with the store that perfectly meets their needs. Especially since they have all the reward points in the world to save them a nice chunk of change.

  Best Places to Shop for a Phone: Phone service providers, electronic stores the offer reward points (e.g. Best Buy, Walmart, Target, etc.)

- **Shopper on a Mission** – You are more of a get in and get out type shopper. Time and budgeting is your most valuable resource. To make shopping easier, you craft your budget all the way down to amount of sales tax before you go shopping. Your goal is to always match or stay under that budget and get out as quickly as possible.

  Best Places to Shop for a Phone: Anywhere or places where there won't have a long wait time.

Different Types of Phones

As stated before, there are a lot of phones on the market today. This section will help your narrow your options down by breaking these many phones down into simple categories. Take a look at the information below to receive some first-hand information of what the phones are really made of.

Conventional Phones

The phone market is steadily shifting over to smartphones, and conventional or regular mobile phones are slowly being pushed out of the way by their descendants. However, prepaid phone services seem to keep a great supply of these phones. These phones are also very inexpensive when compared to the smartphone. The best features that these phones offer is their compactness and the easy to use keypad/keyboard. Many have cameras and support for wireless Bluetooth headsets for hands-free communication. Many can access high-speed data networks to enjoy music and video-based services. Other capabilities might include a touch screen, a QWERTY keyboard, a full browser, a multi-megapixel camera, memory-card storage for music and pictures, and more options for custom ring tones, games, and other services. If you are searching for something functional, then this phone will be the right one for you. The same advice we gave above about form factors also applies to feature phones. In this case, in addition to touchscreens, sliders, and QWERTY slabs, you'll also have basic voice phones that are either candy-bar shaped with numeric keypads, or flip phones that open up to a larger, more comfortable numeric keypad. Flip phones have the added benefit of not needing a keyboard lock; close the phone and you won't mistakenly dial someone while it's in your pocket. For the accident-prone, some are even waterproof or ruggedized.

Smartphones

Their advanced operating systems give them access to a host of applications: productivity tools, shopping, multimedia, games, travel, news, weather, social media, finance, references, etc. Popular, high-profile models such as the Samsung Galaxy and the iPhone command the highest prices. But there are a massive number of older model smartphones that can achieve most of the same functions for about a fraction of the price. These older smartphones perform much better than the conventional phones, and the offer the appeasing accessibility.

Form Factors

The form factor refers to the position and layout of the phone's major elements, shape, size, and style. These form factors are prevalent in flip phones, bar phones, slider phones, their subcategories, and uncommon models.

Bar Phones *(Also known as: Candy Bar, Block, and Slab)*

This phone takes on the shape of a cube and has rounded edges and corners. It vaguely resembles a chocolate candy bar, thus the name bar. The bar is a form factor this is used by a large number of phone manufactures, and they all have a similar layout. All bar phones contain only a few buttons on the outer shell of the phone. This small group of buttons includes a power, volume, and camera button on the outer left and right edges of the phone.

There are a few buttons located at the very bottom of the touchscreen, the home button, the back button, and the search button. The earphone jack is located at the top left or right corner, and charger jack on either the bottom-center, left, or right corner of the phone. If the charger jack is located on the left or right, it is usually located on the side opposite of the power button. Only the QWERTY virtual keyboard is available with this phone. This is the most frequent form factor that you will see in mobile phones today.

Touchscreen *(Also known as: Slate Phone; Subcategory of the Bar Phone)*

The touchscreen is prevalent in smartphones and all advanced mobile devices. It consists of an onscreen QWERTY virtual keyboard. Nine out of ten smartphones worldwide are manufactured in touchscreen form.

Phablet *(Subcategory of the Touchscreen)*

The name *'Phablet'* is a combination of the words phone and tablet. These are the larger smartphones that can function as both a phone and a tablet. The screens of phablet usually measures to about 5" to

6.9".

The Swivel Phone

The swivel phone is made up of two segments that are connected by a central axis which swiveled past each other in a way that usually resembled a kick. . These phones were made to allow the consumers to have a physical keyboard accessible to them without being too large in size. Some later slider phones came equipped with both a touchscreen QWERTY keyboard and a physical QWERTY keyboard.

The Brick Phone

The Brick phone mainly refers to what will now be considered a chunky, passé bar-shaped phone. These phones usually consisted of a very large battery and an alphanumeric keypad. Earlier "bricks" were accompanied by an antenna protruding out of the top of the phone. However, many apply the brick to older touchscreen phones and other form factors like the slider phone and the flip phone.

The Flip Phone *(Also known as: The Clamshell)*

The flip phone consists of two sections that are connected by hinges that allow the phone to open close just like a clamshell. This design allows for a much more compact and portable phone. When open, the top section of the phone exposed a screen along with a speaker, while the bottom exposed a keypad, a few command buttons, and a microphone. The bottom half is where the battery was housed, while outer top housed a digital clock display.

The Slider Phone

The slider phone is a phone that is made up of two or more sections that use a rail system to slide open. Once the phone slid open the bottom half mainly revealed a small QWERTY keyboard and the top half house the phone's screen, microphone, and speaker. These phones were made to allow the consumers to have a physical keyboard accessible to them without being too large in size. Some later slider phones came equipped with both a touchscreen QWERTY keyboard and a physical QWERTY keyboard.

The Taco Phone

The taco phone was very short-lived. The phone resembled a taco when talked into, due to the microphone being located on the side of the phone. The phone consisted of a display screen on the front center, a keypad of the right, and a group of command buttons on the left.

The Watch Phone *(Also known as: Smartwatch)*

This is a wristwatch with Bluetooth support and full 4 band Global System for Mobile Communications (GSM) to allow the user to make a phone call.

Phone Manufacturers

There are many phones on the market today,

Samsung Galaxy

The Samsung Galaxy or Samsung GALAXY is the dominating mobile device for all Android phones. This phone is manufactured, designed, and advertised by Samsung. The Galaxy line of products includes state-of-the-art smartphones, a tablet series, and phablet series. The mobile devices are definitely making a statement, and it's obvious that everyone is paying attention. The Galaxy phones are the runner-up to the iPhone when it comes to the most purchased. However, this phone has had extraordinary success. Due to its selection of over 50 devices, Samsung has been able to break in to all pricing options, every mobile carrier, every country, and every form factor.

The development of the Samsung galaxy began in 2009 with the release of the Samsung i7500 (aka Samsung Galaxy). This was the beginning of Samsung's race into the Android market. Today, the Galaxy is in great competition with the iPhone. Ironically, until recently Samsung had been manufacturing chips and phones for the iPhone. If you plan to purchase this phone, then it would be a great choice due to the level of its accessibility and user – interface. However, be aware of the cost of the phone.

Apple iPhone

The iPhone is a line of smartphones designed and marketed by Apple. The phone was released back in 2007, and has dominated the smartphone market since its release, and it sold approximately 6.1 million iPhone units over five quarters. There is currently eight generations of iPhone, and eight generations of Apple operating systems. The first generation phone was a Global System for Mobile Communications (GSM) phone, some of the designs and physical features of the first generation iPhone have been replaced with better features by the newer models. The newest model of the iPhone has a screen that measure to about 4.7" to 5.5". The earlier versions of the iPhone had smaller screens that measured to about 9 cm.

This phone has amazing capabilities that no other phone has on the market today. Features like a talking personal assistant named *Siri* that helps you get all of your tasks done and interacts with you through this outstanding natural speech voice recognition feature. She can also ask you questions if she needs more information about your commands.

The iPhone has been a well-publicized competition with Samsung and Apple seems to winning by the long haul. And now that Samsung is no longer manufacturing chips and phones for Apple, the competition just got easier for Apple. It will be impossible for Samsung to beat the accessibility of the iPhone. Also, just like the Apple computers, the iPhone was built to last.

Brands of Other Phones

- Acer
- Alcatel
- Allview
- Amazon
- Amoi
- Apple
- Archos
- Asus
- AT&T
- Beneton
- BenQ
- BenQ-Siemens
- BlackBerry
- Blu
- Bosch
- Casio
- CAT
- Celken
- Chea Comm
- Dell
- Emporia
- Ericsson
- Eten
- Fujitsu Siemens
- Garmin-Asus
- Gigsbyte
- Gionee
- Haier
- Hauwei
- Hewlett Packard
- HTC
- Icemobile

- i-mate
- i-mobile
- Innostream
- iNQ
- Jolla
- Karbonn
- Kyocera
- Lava
- Lenovo
- LG
- Maxon
- Maxwest
- Meizu
- Microsoft
- Microwave
- Mitac
- Mitsubitshi
- Modu
- Motorola
- Mwg
- NEC
- Neonode
- NIU
- Nokia
- NVIDIA
  OnePlus
- O_2
- Oppo
- Orange
- Palm
- Panasonic
- Pantech
- Parla
- Phillips
- Plum
- Posh
- Prestigio
- Qtek
- Sagem
- Samsung

- Seiwan
- Sendo
- Sharp
- Siemens
- Sonim
- Sony
- Sony Ericsson
- Spice
- Tel.Me.
- Telit
- Theiraya
- T-Mobile
- Toshiba
- Unnecto
- Vertu
- Verykool
- VIVO
- VK Mobie
- Vodafone
- Wiko
- XCute
- Xiaomi
- XOLO
- Yezz
- Yota
- YU
- ZTE

Basic Phone Accessories that will enhance your Phone Experience

There are some essential phone accessories that you must acquire and maintain to be able to utilize all of the features of your phone and keep it running successfully. Below is a comprehensive list that will give some insight on why these accessories are needed for your phone.

Phone Charger

This is will be the accessory that will come with your phone. The phone charger keeps your battery charged. New mobile phones come with a charger that has a detachable USB cord to allow consumers to connect your phone to desktops, laptops, and other devices. This will allow you to transfer files between your phone and other devices. You can also use the detachable USB to charge your phone through other devices.

The Wireless Charger/Wireless Power Transfer (WPT)

Wireless chargers are the new accessories to the market. The technology involves the transmission of electrical power from a power source to the mobile device without the use of any wire connectivity. This allows the phone to be charged when there are times of inconvenience or when there are dangerous circumstances. The way it works is a transmitter that is connected to a power source sends of power to the phone by electromagnetic fields the surface that will or contains the phone. The power will then be converted into electric power and sent to the phone.

Secure Digital (SD) Card

Most smartphones require an SD card for the ultimate data storage. SD cards are interchangeable between devices and usually come with the mobile phone. SD cards come is different sizes, weights, speed classes, and capacities, and will only fit into matching slots. The size compatible with your mobile device will be the microSD card and measures to about 15 x 11 x 1.4 millimeters and weigh about 0.25

grams. The capacity of microSD cards range from 2GB to 32GB. The microSD card that will come with your phone will have a capacity of 4GB. This card will allow you to store media, contacts, and other applications.

Subscriber Identity Module (SIM) Card

This is an integrated circuit that serves the purpose of securely storing International Mobile Subscriber Identity (IMSI) (used to identify a mobile network) and the related key that identifies and authenticates subscribers the mobile device it is assigned to. These cards are designed to be exchangeable between different mobile devices. For your mobile phone, a SIM card will be assigned to you be the phone carrier to activate phone service.

Auxiliary Cord

This cord will connect your phone to any headphone audio jack that your cord can connect to. You will be able to connect the opposite end into the headphone audio jack in your phone to enjoy music and phone audio through the speakers of whatever you are able to connect it to.

Headsets

The headset is not an essential, but it is a very helpful tool. The most popular headset is the Bluetooth headset. This accessory allows you to talk on your phone wirelessly. This accessory is helpful to use while driving, but many new model cars already come equipped with Bluetooth capabilities.

Phone Case

The phone case protects the phone from damage. This can be purchased through phone carriers and other marketplaces.

The Five Pillars of a Good Smartphone – How To Choose The Right Smartphone

The earlier chapters of this book have given you a thorough introduction to the cell phone and how quickly cell phone technology has developed. Since the original smartphone was introduced, the pace of development accelerated even more. We can look forward to that development trend increasing further in the future. Clearly a smartphone can perform many functions, delivering an incredible amount of utility for its owner. All that utility increases the difficulty of choosing the right smartphone for you.

In this chapter, we will talk about the important points to consider in choosing a smartphone. We referred to these of the five pillars of a good smartphone. These are the key funtions that drive overall performance and owner satisfaction. They are also the drivers of cost. In addition, we are going to talk about your personal wants and needs and how they will influence the selection of your smartphone.

- First, we are going to talk about the size and design of your smartphone. You are going to be handling, looking at and viewing your smartphone countless times a day. The design and size are extremely important. If you cannot hold it and operate it comfortably, or you cannot read it easily you will be perpetually frustrated.

- They say that "patience is a virtue", but our expectations for the performance of the technology is high and our patience is low. We will be talking about processor speed and storage.

These capabilities are very important to the usability of your smartphone. If your GPS app, cannot keep up with the speed of your car, it will not be very useful will it? Once you begin to depend on features like GPS navigation, speed and storage are essential for usability and for safety.

- Our third pillar is storage. You will see that this is essential to your personal satisfaction because it relates to using camera features, video functions, listening to music, playing games and overall storage capacity for the information you wish to retain and recall instantly.

- Pillar number four is dedicated to camera functions. These too, relate to processor speed and storage, but can also drive your budget through the roof. Never the less, your picture and video taking capabilities can make a difference in how much fun and utility you get from your smartphone.

- The last pillar is the budget. Hardly anyone escapes being constrained by the straps of the budget. Unfortunately, this is a tricky area to negotiate and can make or break your plans for entering smartphone nirvana.

Getting Grounded

This is not going to be a quick process but it can be easy and fun if you relax and enjoy it. Selecting a smartphone is like shopping for your clothes. It is very personal and subject to your tastes, habits and fit. You have many things to consider and self-examination is required. Nothing else you can buy has the potential to serve you in as many ways as a smartphone. Having a good smartphone is like holding the world in the palm of your hand.

Get your research hat down from the shelf, because it takes thorough investigation to uncover the facts you need to make the many choices at hand. This will be a great learning experience and a good investment in your time. The smartphone market changes so quickly that you will be able to use the experience when you need to replace your smartphone for newer technology.

Choosing is not a sequential process. You will need to do a little work in each pillar area and then switch back and forth between them until you have balanced all your requirements and fit them into your budget.

So take out your favorite journal or pad and finest writing instrument. Are you ready to take notes? Ok, here we go.

The Preliminaries

Choosing a smartphone needs to be placed within the right context.

If you do not have one already, you will have to choose a wireless network provider such as AT&T, Sprint, T-Mobile, or Verizon. This book is not about choosing a network provider, but it is important, so here are a couple of essential points.

Understand what you need and make the best choice for you. There are three main focal points.

- COVERAGE: Your smartphone needs to <u>work well</u> from home, work, where you hang out, on the highways and byways you use. If you travel a lot, you need to check out how well those destinations are covered.

- EQUIPMENT: Can you buy your smartphone elsewhere, or do you have to buy it from the provider. What is the difference in support and price based on where you buy it?

- PRICING: Understand what components affect the price and understand what your needs are. Do not buy more than you need. You can upgrade later if you need to.

Do your research with the help of knowledgeable sources. Use the Internet to dig for information. There are organizations that publish evaluations and comparisons annually. Check these for starters but searching the internet will get you many more.

WWW.CNET.COM

WWW.PCWORLD.COM

WWW.TECHREPUBLIC.COM

If you do have a wireless network provider already, and you are happy with that provider, then you only need to address equipment and pricing.

The big network providers have the broadest choices for smartphones. The local providers have narrower offerings.

Size and Design

Let me start with size because one size does not fit all. You have to get this right. Here is what is important.

- It must be comfortable in the palm of your hand. This size and shape of your hand and fingers is a big influence on comfort and ease of operation. I have small hands with fat fingers, what do you have? This is a very important item. The only way to evaluate this is a field trip to the supplier, where you can pick it up and handle it every conceivable way. Do not overlook weight and balance. Hold it up to your ear. Pretend to be on a call. How long can you hold it up there? Can you feel cramps developing in your hand? Will you need to operate it with one hand? Try that out. Do you do a lot of texting? Try using both hands on the keyboard simultaneously. How does that change the feel of balance and comfort?

- Where are you going to carry it? If the answer to that question is "in my purse", your pretty much done with this concern. Otherwise, you had better figure out what to wear or bring on your field trip. Many smartphones will not fit well in pockets due to large screen sizes. Continue your tests. Put the device where you will carry it. Pretend you are receiving a call or a text and retrieve the device. How does it handle when you are taking it out of your carrying location. How easy is it to put it back?

- How big is that screen anyway? Do you actually need a big screen? Devices with large screens can be unwieldy and difficult to store. Also, you cannot effectively replace your TV with a smartphone. So think about what is driving you to want a large screen. Will you watch video on the screen for long periods? Do you have a vision problem? Does your choice of phone support visually impaired display functions? How about the on-screen keyboard? Try it out, especially if you have fat fingers. The keys may look larger, but do they work any better when you try to use them? Fat finger alternatives are stylus pens and voice commands. Those might be better for you and eliminate the higher cost of a larger screen device. With all the graphics and video used today and planned for the future, you need a reasonably sized screen. Most people can live happily with between 4.7" and 5" screen, but never go less than 3.5".

Let us talk about the quality of images on the screen. There is more hype around this than anyone would really care to encounter. What counts is that the screen works well for you. Forget about the technology. Many people have vision problems and are unable to discern subtle differences in image quality. On your field trip, you need to test the phones and compare the screens. You will need sales assistance for this test.

Ask how to put the device in Auto Mode, used for saving battery. Now put the phone through some paces. Look at web pages, pictures, text, video and stare at the screen closely. Do you notice any of the following?

- Reflection

- Dull or artificial looking colors

- Blurry text

- Feeling of eyestrain

If you do not experience any of these, then you are in good shape. Otherwise, keep shopping or consider the need to get an eye exam.

Moving on to design, most designs are for use in friendly environments. Will you be using your smartphone in tough environments? Might it be subject to drops or moisture? You may want to look a ruggedized design. That changes dimensions and weight significantly. However, if that is what you need to get utility from your phone, then that is the way you have to go. Sorry!

Keyboards are important. The touch screens on most smartphones have multiple keyboard style options. However, if you are a large person, you may have trouble with on-screen keyboards. If you cannot interact easily with your device, you will be unhappy. Start looking at tactile keyboard devices immediately. Some manufacturers have done a nice job for folks with large hands and fingers.

Controls for things like volume, should be located on the sides of the device and easily manipulated with one hand. This is another test item for your field trip. Have fun with it. Pretend your significant other or boss is screaming at you over the phone. How quickly and easily can you get the volume down?

Attaching things like earphones and chargers should be easy and when connected, should not hamper your operation of the device. Be sure to check while on your field trip.

Let us not forget appearance. Some smartphones have metal skins and others are plastic. Finishes can be shiny, dull, or textured. They can come in nifty color schemes or just boring black. This might not matter to you if you intend to apply an external case to your phone. Cases are relatively inexpensive and provide a little extra protection. You can put a personal touch on some cases by using your own photos and graphics. If you are interested in this, it will be an added cost, though a small one.

Processor Speed and RAM

A smartphone is actually a computer that can provide telecommunications functions. It is very sophisticated and manages many tasks simultaneously. It can store huge quantities of information within its tiny framework, and needs to be able to retrieve it instantly. It also manages real-time communications over multiple networks, for data, the Internet, voice, video, and positioning. This means your smartphone is running many programs simultaneously.

The number and type of programs you run drives the need for processing power and the number of processors. Each running program is stored in RAM (random access memory) to make it easy for the processor to execute the instructions.

There are millions of applications or APPS you can acquire to run on your smartphone as well. Some of these can place a very large burden on your device in terms of processor power and RAM. For example, a GPS navigation app runs multiple programs that work together. It can be a blockbuster on processor power and RAM. If you were to be using this application frequently, you might want to be sure to have multiple processors.

Your smartphone's basic design has the processor configuration and RAM it needs to provide high performance for all its inherent functions plus an allowance for running a reasonable number of apps. How you personally use your device, will dictate the need for more or less processing power and RAM. If you keep many programs running simultaneously, you will need more processor power and RAM to ensure you get high performance at all times. If you are the less demanding kind of user, you will need less.

Once you become a smartphone user, you will notice that apps are updated frequently to increase functionality and correct errors. As with PCs and larger computers, app design holds little regard for your devices resources. That means the longer you own your device, the more updates you will receive and the more processor and RAM capacity is used. This will likely result in you finding it gets slower over time. To avoid this, consider the following guideline.

- If you feel like you are going to be happy with the standard features of your device and only very few apps, do not worry about the processor speed or RAM. You will be fine.

- If you think owning this device is going to be fun and you are looking forward to all the dynamite apps you can use, consider a 1GHz processor and minimum 512MB of RAM package.

Storage – Internal and External

Storage is a tricky subject. Again, it is going to depend on how you use your device. All devices come with system memory. This is the internal storage. A large portion of that goes to operate the device, but the rest is available for apps and data. Right now, many smartphones have options for 16, 32 or 64 GB of system memory. If your device uses 6GB to run itself, subtract that from the total and that is what you will have left for everything else.

The incremental price of those system memory options is high. An alternative is to find a device that permits removable storage such as an SD Card. This is external storage. However, do not get confused by the term. The card is actually stored within the device. These cards come in similar sizes to system memory but are less expensive. They are great for storing music, pictures and video that take up large amounts of memory. Buying a device with 16GB of system memory that also supports the use of an SD card, should be less expensive than buying a device with more system memory. Do the math to be sure.

Another kind of external storage is available through apps that specialize in storage. Storage apps keep your information in the cloud, and not on your device. There are two big things to remember about using these apps.

- Your data cannot be stored or retrieved unless you have a connection to the Internet.

- Large data files such as pictures, videos, and music may take a little longer to access than those stored in system memory or on SD cards.

Storage apps are very inexpensive. Anywhere from free to a few dollars a month. They are a great alternative for storing many types of data.

Cameras

Just about every phone has a camera. Smartphones have what I call serious cameras. That is to say, they can rival the capabilities of some high-end digital standalone cameras. This makes talking about this subject a bit cumbersome because digital photography is a broad technology.

If you do not care, much about taking pictures, or only want to take some occasional shots and selfies, you can jump to the next heading. I guaranty anything you need to photograph with your smartphone will come out just fine. Unbelievably, regardless of megapixel rating, any smartphone picture will look great on any other smartphone or computer display.

Before we go further, let us get something important on the table. Do not make photography functionality the primary criteria for selecting a smartphone. They cannot replace sophisticated digital photography equipment. You will end up being disappointed.

Therefore, you shutterbugs, videophiles and semi-pro photography enthusiasts will need to do homework. Start by writing down what you want to do with the camera in your smartphone, how frequently you will do it and in what quantity. Do not forget to note the conditions you expect to be under, such as high light, low light, reflectivity, indoors, outdoors. Are you going to shoot action stills, videos or whatever? Will you be making prints? What sizes, etc. The more you want, the more you will pay. Here are some things to think about as you do your research.

- Camera functionality can change the dimensions of the smartphone. Most notably, making it thicker and perhaps having a bulge somewhere. This can alter the utility of the smartphone. It may change the way you carry it around with you. It may create an uncomfortable feel due to size and balance.

- The more sophistication built into the camera

functions, the more likely that buttons on the device will take on multiple roles. You must get into this in some detail while on your field trip to the smartphone supplier. It is important to understand any difficulty related to the position and use of the buttons. For example, if you miss the button, will you accidentally power off the device, causing you to miss the shot. Another concern might be the tendency to block the lens with your finger, forcing you to hold the device in an uncomfortable position or ruin the shot.

- Camera activation is another operational characteristic you will want to test. How do you actually activate the camera, when in lock mode or unlock mode? How long does it take?

- Other time related actions can make a difference to your personal satisfaction. How long will it take to get your image in focus in auto and manual modes? How quickly does the zoom feature work? How long does it take to change shooting modes, like changing to panorama?

- Some camera features need to be invoked using on-screen buttons. Try these out paying particular attention to whether you can use them easily and without having to look at the screen.

- Test the camera's image quality at various distances. Some cameras will be better close up, or when the subject is at a longer distance. How will you be using the camera and will you get the quality you want for the most frequently shot subjects.

Here are a few more memory joggers to help stimulate

your research.

- Larger image sensors are better at light collection.
- Regardless of sensor size, squeezing in too many pixels can make grainy images.
- If prints are important, more megapixels yields larger print sizes.
- Do you want flash capability?
- Do you need auto-focus or image stabilization?
- Do not forget to check shutter speed and frames per second regarding video.

The Internet is your friend for this kind of research. Check manufacturer sites and YouTube. Just don't forget that field testing your alternatives is the most import part of the job.

The Internet is your friend for this kind of research. Check manufacturer sites and YouTube.

Price

Smartphones are telecommunications equipment and require a wireless network services to function fully. Network service providers are the kings of razzle-dazzle pricing. Keep in mind that their objective is to keep you as a customer forever. There are no exceptions to this rule.

Smartphones are expensive and their price is justified based on what they do. The capabilities are very enticing and can get you hooked easily. The question is whether you can afford to have what you want.

In reality, smartphones are luxury items but promoted as indispensable by the telecommunications industry. Let me give you some real insight.

On the day I researched this information, you could buy a particular HTC smartphone at a big electronics store, without any wireless network provider services for $599.99. This is not a top of the line phone but a good one. This price is the list price, not discounted at all. Is not the electronics store a discounter? It is because smartphones receive discounts when packaged with wireless network provider services. It is a standard practice in the telecommunications industry. IMPORTANT SIDEBAR: The electronics store is also a re-seller of wireless network provider services. This influences how he sells smartphones. There are other sources for buying smartphones, where the seller does discount the smartphone because he has no affiliation with the wireless network service providers. CONTINUING: Now if you were to choose to purchase this smartphone in conjunction with a wireless network provider service, you would receive a substantial discount. For example, if you agree to take a 30-month network provider service contract, you could own this phone after 30 installment payments of $14.97 for a total of $449. To sweeten this deal, you can upgrade your phone after the first 24 months, at which point you stop paying for this phone and save six payments equal to about $90. Now the total cost of the phone is only $359.00. The plan also waives activation fees and upgrade fees you would have to pay if you supplied your own phone. Why would they do that? To ensure you renew your wireless network service at the end of the contract.

The above is only one example of a pricing plan. There are many more. None of them is tricky, but the marketing puts emphasis on those things that will draw you in, not inform you of the important details. Be diligent and make sure you understand everything.

Getting back to your budget, you are going to need to cover one-time charges and monthly charges regardless of how you go about acquiring your smartphone. Do not forget your smartphone requires a wireless network service provider.

There will be no escaping the need to conduct thoughtful mathematical analysis as you make your choices. Using a spreadsheet will make this easier. Additionally, carefully examine the fine print of the agreements. Cancelation charges will apply if you need to get out of your contracts. To be thorough, you may want to calculate what cancelation can cost you. It might help reinforce the need to make careful decisions.

Your mission

Your selection of a smartphone is a small project that can facilitate making many positive changes in life. Have fun with your project but be practical. Coming up with a budget that is comfortable for you is most important.

Smartphone technology changes quickly. The simpler your needs are, the longer your smartphone will provide good service for you.

Advances in app development are likely to entice you to add more functionality to your device as the months pass. Remember to create an allowance for those additions.

The five pillars of a good smartphone will guide you in looking at your needs and fulfilling them in a stepwise fashion. It will also help you create a balanced approach to controlling the cost of your smartphone. You can work out trade-offs within each pillar and then among the pillars. Do not let the volume of options and considerations take your eye off the real needs you have.

- Size and Design are very important. This is a tool you must handle and take with you all the time. It needs to compliment your body, mind, and spirit.

- <u>Processor and RAM</u> work together to ensure top performance. The more you do and the sophistication of what you do are the influences here.

- <u>Storage</u> is a big variable. If you keep lots of photos, videos, and music for instant access, this can be expensive. Ask yourself if you really need those things instantly available all the time.

- <u>Cameras</u> add fun to your life. This will likely be the first time you have a camera available at every moment. You can capture a lot of memories and information. Do you want to create art, or just record memories? It makes a difference.

- <u>Price</u> is what it comes down to in the end. There are many options and you will need to look at all of them. Be sure to do the math.

A Plan that Works! - Everything You Need To Know About Cell Phone Plans

Choosing a Cellphone Plan

Choosing a phone service provider can be quite difficult from the beginning stages; especially, if you don't know the entirety of your options. There are only five wireless providers that are widely advertised, and even less that offer nationwide coverage. Although these phone carriers have superb service and great coverage, there are some phone carriers that offer dependable service. It is best to shop around and research. However, this book has everything for you in one place. The overview of phone plans in this section will give you great information on the options that you have available. You'll be surprised that your options are slightly wider than you think.

Contract Phone Plans

Contract phone services are the most constricted. They usually offer great coverage and great phones. However, the minimum duration of the contract is two years. So, if you are not able to make the payments on time consistently, your credit will take a major hit. And unfortunately, for individuals with bad credit, it will be hard to receive a phone under a contract. The only thing that makes contract phones enticing is the drastic reduction of the phone prices. Purchasing a phone under contract will reduce the prices of some phones all the way down to FREE! Some phones can be reduced by $100.00 to $400.00. Most consumers with a contracted phone plan have to wait to upgrade their phones. It can be considered best to avoid any obligations to pay a monthly bill. So, don't be seduced by the lowered phone prices of the phone. This will be the ideal phone plan for the mission shopper or the researcher, but not the impulsive shopper. There are a number of phone carriers that offer contract phone plans. Check out the list below:

- AT&T

- CREDO Mobile

- Sprint

- T-Mobile

- US Cellular

- Verizon Wireless

Prepaid Phone Plans

Prepaid phone plans offer you the option of paying a monthly fee with no contract. This option works great for those that are not financially able to meet the obligations of a contract, or risk their credit taking a hit. Another upside to this phone service is that when it comes to upgrades, the sky is the limit. You can BYOP (Bring your own phone), or you can upgrade or downgrade to any phone that you like, at any time you like. However, this comes at a price. You will have to pay full price for the phone if there is no contract. If you aren't prepared to do this, then it is best not hold-off on this plan until you have the money to accommodate. Plan ahead for the best phone that you can afford, and if you do enough saving you will be able to purchase the phone that you want. For this phone plan, it would help to adapt to the characteristics of the researcher, the discount shopper, and the negotiator. Impulsive buyers stay away unless you are prepared to pay. Here is a list of phone carriers that offer prepaid phone services.

- AT&T
- Boost Mobile
- Cricket Wireless
- Freedom Pop
- Freedom Pop
- GiV Mobile
- Go Phone
- H_2O Wireless
- Metro PCS
- Net 10
- Page Plus Cellular
- Pay-Lo Talk & Text

- Republic Wireless

- ROK Mobile

- Simple Mobile

- Sprint

- Straight Talk

- Ting Mobile

- TracFone Wireless

- US Cellular

- US Mobile

- Verizon Wireless

- Virgin Mobile

- Walmart Family Mobile

- Zact

Pay-As-You-Go Phone Service

This phone Pay-As-You-Go phone plan is another option that does not require a contract. With this plan, you have the option available to pay only for the talking, texting, and data you use. You just refresh your minutes and data whenever they run out. There is no monthly fee, and you are not obligated. You have the same freedom as you would under a prepaid phone plan. This plan also requires that you pay for your own phone or bring your own phone. However, some phone carriers will require a SIM card that you will have to purchase if you want to use their mobile services. Here are a few Pay-As-You-Go service providers below.

- AT&T
- Boost Mobile
- Consumer Cellular
- Earthtones
- Jitterbug
- Kajeet
- Liberty Wireless
- Metro PCS
- Net 10
- o2 Wireless
- Page Plus
- PTei
- Ready Mobile
- Red Pocket Mobile
- Republic Wireless
- Simple Mobile

- Solovei

- Sprint

- STI Mobile

- Straight Talk

- T-Mobile

- Total Call Mobile

- Touch Mobile

- TracFone Wireless

- US Cellular

- Verizon Wireless

- Virgin Mobile

Wi-Fi Phone Plans

Wi-Fi phone services are sort of new to the phone carrier market. The offer no contract carrier-free, Pay-As-You-Go phone service. There is no need to go to a store, because these plans are all enabled through mobile applications. These apps offer free texting and some offer a certain amount of free talking minutes. However, you will be able to purchase a large amount of minutes at a very low price. For example, some applications offer 1000 talk minutes for as low as $14.99 per month, with free talking, and data is free because you will be running off of Wi-Fi. Some of these phone services will even allow you to retrieve free minutes by completing phone surveys and trying out new apps. This phone service may be unconventional, but it is a money saver, and can be very helpful to those that are able to adapt. Here are a few Wi-Fi based phone services.

- KiK
- Line
- Pinger
- Ring Plus
- Skype
- Text Free
- Text Now
- Text Plus
- TextMe
- Walkee Textee
- WhatsApp

Taps, Clicks, Troubleshooting, and Phone Tricks

This chapter is all about what hidden features your phone can perform, and some of the basic troubleshooting that you can do when your phone is not working the best for you. However, your best resource will be your phone manual.

Mobile Phone Signal

The strength of your phone signal has a major influence on the functionality of your phone. Failed text messages, poor call quality, or slow running applications usually indicate that there is a problem with your phone service. If you run into this problem, review the information below to assist you with troubleshooting.

Full Reception/Full Bars

If the reception bars on the phone are high, this indicates a very strong signal. If you are having difficulties placing calls, sending messages, or using other applications, but your phone indicates a strong signal, follow these steps to troubleshoot:

Restart your phone – Most problems like this can be resolved by simply restarting your phone. There are two ways that you can restart your phone:

Remove the Battery

You may restart your phone by removing the battery and then immediately putting the battery back into the phone. Once you have put the battery back into the phone, you will then power the phone back on. Once the phone is powered on, confirm a signal and attempt to send the text message again. For the best result, wait about 5 minutes before powering the phone back on.

Power Off/Power On

You may also restart your phone simply powering the phone off, and then immediately powering the phone back on. Once the phone is powered on, confirm a signal and attempt to send the text message again. For the best result, wait about 5 minutes before powering the phone back on.

If the problem persists, you will need to check with your mobile carrier to have the problem fixed immediately.

Weak Signal/Low Bars

If the reception bars on the phone are low, this indicates a weak signal. If you are having difficulties placing calls, sending messages, or using other applications, and your phone indicates a strong signal, follow these steps to troubleshoot:

- **Clear Obstructions** – If your phone signal has weakened, you will need to check for possible obstructions (location, trees, tall buildings, etc.) in the area. It's possible that you are not in close enough range to a cell phone tower, or there may an object blocking the signal. These are the most common causes of a failed text message and weakened signal. Once you have determined and solved the problem, and the signal is restored, you should and be able to send the text message.

- **Restart your phone** – Most problems with low reception by restarting your phone. There are two ways that you can restart your phone:

Remove the Battery

Click here to be redirected to the instructions above.

Power Off/Power On

Click here to be redirected to the instructions above.

If the problem persists, you will need to check with your mobile carrier to have the problem fixed immediately.

No Service/No Signal

If the reception bars are replaced with the prompt, 'NO SERVICE' (indicates no signal), you may need to restart your phone or even remove the battery. Once the phone has powered back on, check to see if the signal has been restored. If there is still no signal, you will need to check with your mobile carrier to have the problem fixed immediately.

- **Check (IMEI)** – Sometimes consumers will be sold a phone that has a bad IMEI. This usually means that your phone has been blacklisted in order to prevent use of the phone due to the chance of it being a stolen phone. Even though your phone is not stolen, this could still happen in error. You can check the IMEI of your phone and unlock it through the internet. There are many websites that offer this service available online. If you are still experiencing issue with your phone, you will need to talk to the phone carrier and the store that you purchased the phone from.

- **Clear Obstructions** – *Click here to be redirected to the instructions above.*

- **Restart your phone** – Most problems with low reception by restarting your phone. There are two ways that you can restart your phone:

Remove the Battery

Click here to be redirected to the instructions above.

Power Off/Power On

Click here to be redirected to the instructions above.

If the problem persists, you will need to check with your mobile carrier to have the problem fixed immediately.

Cool Features

Your phone has some very cool feature many people are not keen to. This chapter will put you up to date on some of the hidden features of your phone. Most people only become aware of these features after they've had their phone for a while. If only they would just read the phone manual. But who really reads those anyway? Luckily, this book is here to give you a speed pass through the essentials of what your phone can do, and some of the cool things that make your phone

Take a screenshot

A screenshot is an image captured by your phone to record the items displayed on the monitor, the LCD of your touchscreen phone. Normally, this image can be taken using a mobile application that you can download on your phone. However, there is no need to do that because your phone already has the capability to capture a screenshot. Take a look below at how to take a screenshot with the most prevalent phones in the industry.

Iphone

To Take a Screenshot with the iPhone:

Press and hold down the 'HOME' button simultaneously with the 'SLEEP/WAKE' button. Once you hear the camera shutter, you have successfully taken a screenshot with your mobile phone. The phone will automatically save to your photo gallery/photos.

Android

To Take a Screenshot with the Android:

Press and hold down the 'POWER' and 'VOLUME DOWN' buttons simultaneously. Once you hear the camera shutter, you have successfully taken a screenshot with your mobile phone. The phone will automatically save to your photo gallery/photos.

Windows Phone

To Take a Screenshot with the Android:

Press and hold down the 'HOME' and 'POWER' buttons simultaneously. Once you hear the camera shutter, you have successfully taken a screenshot with your mobile phone. The phone will automatically save to your photo gallery/photos.

<u>Your Phone Can Read!</u>

Find yourself multitasking a lot, this feature will help you manage your multitasking efficiently. The Voice over application allows your phone to read what is available on the screen of your phone at the time. Really comes in handy when you are using your navigation application, or when you receive a text message while driving. This feature will help you keep your eyes on the road.

iPhone

Go to the 'SETTINGS' menu, and select 'GENERAL'. Then select 'ACCESSIBILITY'. Next, enable the 'Voice Over' option.

Note: This may take some time for you to get used to. So, feel free to practice before you actually begin using this feature regularly.

Android

Go to the 'SETTINGS' menu. Then select 'ACCESSIBILITY'. Next, enable the 'TALK BACK' option.

Note: You can adjust your Talk Back settings by selecting the Text-to-Speech application through the accessibility menu. Here you can control the speed and volume of your phone's speech.

Conclusion

Hopefully my book, *The Smartphone Buyer's Guide 2015*, has provided you with all the information you need to help you purchase the perfect phone for YOU! There is an endless number of ways that you can use technology to communicate with others. However, your mobile phone trumps them all. It can do everything those other devices just can't do, and even everything that they can. The advancements of wireless technology will allow you get the most out of your new mobile device, no matter how new or how old the phone is.

The information contained in this book is detailed and extensive, but it has been developed to give you the easiest shopping experience possible. There are some very important things to keep in mind. You must know what your phone is capable of as well as know what you will need your phone to do for you. The knowledge you gain from reading this book will help you get the MOST out of your mobile phone. Always keep your budget in mind when purchasing anything, and more importantly your phone.

Thank you for reading "The Smartphone Buyer's Guide 2015". If you enjoyed this book, please take the time to share your thoughts and consider leaving a review at Amazon, even if it's only a line or two; it would make all the difference and would be greatly appreciated.

Thank you   *Steve Johnson*

www.ingramcontent.com/pod-product-compliance
Lightning Source LLC
Chambersburg PA
CBHW070824180526
45168CB00002B/738